How to succeed as a Process Professional

I0475462

A Practitioner's Guide

by

Aditi Chopra

DEDICATION

I want to dedicate this book to my mentors who saw the
potential in me and helped me excel in the business
process field.

Preface

I started my journey in the process profession with no process background or formal education. Instead I had a solid software engineering experience.

Software engineering is an interesting field and I spent several years architecting, designing and programming networking protocols. It was certainly a rewarding experience specially when I saw my software code being used at customer sites.

While working as a software development manager, I started seeing software development process in a completely different light. As a manager, I saw where the deficiencies were in the software development process. I also realized why I as an engineer was at times handicapped because of those deficiencies.

As a manager, I was in a position to see the big picture and the gaps that existed in the overall process. I realized that these gaps could easily be filled and that is how I started my career in the process profession.

I learnt a lot on the job and also succeeded in numerous process initiatives being a natural at it. Most of what I did came naturally to me and I was simply driven by the need to make improvements. I could see the gaps and solutions much before other people in the business did and then my job was to make others realize those gaps and need for solutions. That is how I became a change agent.

It is very interesting for me to look back and realize how natural this journey was for me. My supervisors seeing this inherent talent in me gave me several opportunities to grow in the field. At times, I was overburdened with numerous tasks but I gained a lot of experience very quickly by doing those tasks.

I was motivated to write this book so I could share my practical experience with all those who want to succeed in the process profession. I have had people ask me "What kind of training should I take to become a change agent?" My answer to them is first ask yourself "Is this what I want to do?", "Is this something that comes naturally to me?" Asking these questions is very important. The answers will help you decide if this profession is for you or not. Realizing that it comes naturally to you makes it so much easier to succeed. Otherwise it will be a long battle.

Aditi Chopra

CONTENTS

1 SCIENCE BEHIND A PROCESS

In my viewpoint all Process Professionals are organized and methodical individuals. They like methodical surroundings and like to keep things around them in order. They like to follow a certain procedure and go about it an organized way. This is pretty much the definition of process in a nutshell.

Every time a new process is defined, there is a strong need for it! **A need to bring order amidst chaos.** Keeping that goal in mind, process professional analyzes the situation and then formulates a process that meets the need of the moment. This process is then evangelized and adopted throughout an organization and an order is established.

What really is the secret of a successful process? Why would one process succeed and another doesn't? The answer is simple, **it is when a process solves a business problem, it is a success!**

If a process is instituted to comply to standards or to simply get a certification, it is no-op. Such processes are easily forgotten and not followed.

The biggest example of this is **Six Sigma** which originated in Motorola to solve their business problems. It later was adopted by General Electric and helped them save millions of dollars as well as provide efficiency in their business processes.

Six Sigma is one of the most popular process models used in various industries. The key to its success is its focus in solving business problems. Many a times, organizations create process to get certifications. Creating a process is not enough, you need to ask the question "What business problem am I solving by creating this process?" Put it a little differently, if you are indeed creating a process to get certification, might as well use it solve a business problem.

Process Improvement

You might wonder what is the need for Process Improvement? Why not formulate a good enough process to start with? Why not think of all the aspects of the process in the beginning so we don't have to improve it? These are all valid questions!

Going back to the original question, so then Why Improve Process? Following are some of the reasons why:

Process Context has changed

If a process was established long time back and context of the process has now changed, it requires a change in the process as well. Process context can change in various ways. This happens when the organization following it has changed or the technology associated with the process has changed. All these changes beg for a Process Improvement.

An example is lets say a process was established to collect metrics from a defect tracking system. The defect tracking system is then being replaced with a more robust system, this will require that the process used to collect metrics from defect tracking system be changed as well.

In an organization I was working with, a process was put in place to track code review comments gathered during software code review. As part of this process, a tool was used to collect all of the review comments from the code review.

This process became very popular in the organization and management decided to extend the code review process to reviewing design documents. Now since the nature of code and design document review is different, it would require that some tweaks be made in the review process as well as tool to accommodate design document reviews.

Process Soul needs to be restored

As an outsider what we see sometimes is that the soul of the process has been distorted or changed or worse forgotten! This can happen if the process creator has moved on and people responsible for the process maintenance follow a different philosophy.

Process Improvement is thus needed to restore the soul of the process. So what exactly is the soul of the process? Soul is nothing but the purpose behind the process creation. What was the business problem that this process was trying to solve?What goals did the process creator had in mind when creating the process in the beginning?

Is the process still achieving those goals and serving that business purpose? If not, an improvement is required to restore its soul.

Some examples of these can be seen in the service industry where processes are established to serve the clients or customers. Service industry needs to be very customer driven. If the customers are served well, they are happy and business flourishes. A service organization usually starts small and has a well defined process where employees are trained on serving the customers to their highest satisfaction.

As the organization grows and becomes a giant in size, if adequate leadership and a well defined process that scales is not in place, it may lead to diminishing customer satisfaction. This may also lead to senior experienced employees leaving the organization as they are no longer happy with the culture as they were used to.

If and when this happens, recognizing it and taking action to restore the soul of the process is very necessary. Leaders play a strong role in recognizing this situation and then they assemble a team of experts to make the necessary changes.

Process Scale has changed

Each process has a business goal associated with it. As we expand a process or it gets adopted by a wider organization, similar procedures that worked earlier may not satisfy the original goal of the process. Therefore process needs to be improved to scale to the new requirements.

If a process starts small in a business unit and it becomes successful, word often gets out and other business units in the organization want to adopt it as well. However all business units don't follow the same ways of conducting business, their culture might differ.

In this age of mergers and acquisitions, often companies that are acquired follow a different culture, tools and procedures. In such cases, the process needs to be improved or modified to adopt to the new culture of ways of conducting business.

These are some of the reasons why a process needs to be improved. Process change or improvement can sometimes be more difficult than defining a new process. When we are defining a new process, there is no benchmark and it is easier for people to adopt a new process. When there are no procedures established, putting in a process is simply communicating it to the required audience.

However if a process already exists, making a change or improvement requires process followers to undo previous habits and acquire new habits. More will be discussed about this subject in later chapters.

2 TECHNIQUES OF PROCESS IMPROVEMENT

We understand why and when we need to improve a process. Let us now look at how to improve a process and what techniques can we use.

Let's explore some of the approaches one could take in process improvement field. Most importantly, one needs to understand that one size doesn't fit all. Process engineering is a creative art and solutions are unique to the situation and the business problem that an organization is facing. Each situation will have its own nuances and different personalities to deal with in the organization.

Having said that, here are a few strategic approaches to process improvement that can be tailored for a specific need.

Streamlining and removing duplication

Often times, duplication is seen in a process. A streamlining is then necessary to redefine the existing process.

This technique has several advantages. In general, removing duplication leads to efficiency and makes the process more productive.

Lets look at an example. When a process specialist realized that a company was using thirteen different tools to survey their customers, he saw an opportunity to streamline the survey process and removing duplication.

Many business units had the need for surveying the customers but there was no central unit monitoring or guiding the process. Each business unit was using a different process and tool to survey the customers. There was a lot of cost associated with all the different tools and the process and training associated with them. There was chaos all around, people didn't know which tool was better and why they should use one tool over another. A team was then put together to look at the pros and cons of different survey tools and various scenarios in which surveys were done. This team then collectively decided to shortlist two survey tools and get rid of the other redundant tools.

These two shortlisted tools served the purpose of different organizations in the company.

Re-engineering approach, complete re-design

If a process has been around for several years and the context around it has changed to a large extent, then a complete redesign is warranted. It is better to redesign if the current process is not effective and is not solving business problem.

This approach is revolutionary and rather difficult to implement. It takes time and also one needs to deal with opposition and resistance from those who are practicing the current process. But this is indeed a necessity in certain circumstances.

Software companies often redesign their software code if the scaling requirements or performance requirements change and the software module can no longer scale to the new requirements. Similarly a process needs to be redesigned if it needs to scale to a different level.

A process that was working for a smaller business unit will not scale to a larger organization if it wasn't designed to work for larger organizations. In this case a whiteboard discussion is warranted to re-design the process considering the larger scale in mind.

In one of the companies I was working for, quality team had established a process for root cause analysis of software defects. The process was based on the defect tracking system. There were pre-defined fields in the defect tracking system that were used to collect data for root cause analysis. When this process was instituted, it felt right and was the only option at the time given the limitation of the defect tracking system. However, some teams realized the limitation of the pre-defined fields and they couldn't use these fields to do in-depth root cause analysis.

At a later time, this company decided to replace the old defect tracking system with a more robust system which provided more flexibility in terms of data collection. Quality team saw this as an opportunity to enhance the root cause analysis process. They took advantage of this new robust defect tracking system and decided to re-engineer the root cause analysis system.

They decided to add more free form fields to collect more in-depth data which would aid in the root cause analysis of software defects.

There was obviously opposition from teams who were used to using the previous root cause analysis system since they had a lot of data that would now become impossible to use. These objections were taken into consideration when re-designing the system such that old data can be mapped to newer data and not let it go to waste.

Re-designing is inevitable in certain circumstances, however it should be done seamlessly such that older valuable data can still be utilized and mapped to the newer process.

Benchmarking - looking at the industry

Another technique to improving process is looking at the competition in the industry and learning from the competition and using that as a benchmark to make improvements.

Similar to companies doing a competitive analysis on products, a process competitive analysis can be done to learn from industry experience.

Let take a look at a good example of this approach. Lets say a business unit wants to get TL9000 certification done on its projects. This business unit doesn't know where to start and how much process should it institute to get TL9000 certification. Instead of starting from scratch, it can learn from another business unit that is already TL9000 certified. This has two advantages, it speeds up the process and business unit can avoid making similar mistakes and learn from other business unit experiences.

Idealizing approach - copy ideal processes

While researching, a process professional may come across several ideal processes that are already adopted by other organizations. If these processes are proven to be effective, it may not be a bad idea to simply copy these processes.

Why re-invent the wheel if you don't need to? In fact, it might be easier to convince people to adopt an already established process because there are already results associated with it. This will make overcoming resistance easier for sure.

If a process is already established and yielding good results in terms of solving a business problem, then one can easily copy the ideal process.

Of course a lot of time needs to be spent to understand if the context of the process is similar or not but once that determination is done, this is a sure shot way to succeed in the process improvement.

This approach is easiest and best suited in certain scenarios. Only thing you have to overcome as a process professional is your ego. If you are one of those individuals who thinks "If it ain't discovered by me, it ain't good", then this technique won't work for you. However, if you are flexible and wise, and this technique fits well in your scenario, it can work wonders.

I was once working for an organization where I was asked to come up with a set of metrics for an entire organization. Luckily I came across a smaller subunit within this organization that had implemented similar metrics and had wonderful results associated with it.

That metric was just perfect for the business problem that I was asked to solve. I took that opportunity and started evangelizing the results of existing metric to the bigger organization. Pretty soon, they saw the results and the advantage of the metric and the entire organization was ready to adopt it.

3 WHY PROCESS IMPROVEMENT INITIATIVES FAIL?

You might wonder why some process improvement initiatives fail? What are the most common reasons why process professionals are able to launch the rocket but it doesn't reach its destination? Before launching your process improvement initiative, if you want your process initiatives to succeed, you need to be aware of the following reasons of failure.

Inadequate management/leadership support

The top reason of failure is lack of or inadequate management/leadership support. Executive management support is the key to success of any process improvement initiative.

Without management support, an initiative cannot further beyond a concept. Management has to be fully bought into the concept and how the concept is going to help solve the business problem.

If management has not bought into the concept and willing to sponsor it all the way through, that means it is not the right time to launch it. If you feel that management is not fully committed, wait until you have found the right executive sponsor and support from management.

If the leader you are speaking to is not convinced of the merit of the initiative or doesn't fully understand the implications of the business problem, his heart is not fully into the initiative. If such is the case, he may sign up to be sponsor but he might drop the initiative at a later point of time. Perhaps in that case, you need to begin your search for another sponsor or executive leader who fully understands the problem and the merits of the initiative. Don't opt for a false start even though it might feel like a win, it will eventually not work.

Lack of continuous improvement

Another reason for failure could be lack of continuous improvement. Once the process is launched successfully and management support is also established, a close monitoring of the process still needs to happen.

Process control is equally important as is launching the process initiative. When proper procedures are established to keep the process in check and control, it strengthens the initiative. Control procedures provide feedback into the practice of the process and help identify any gaps.

Some changes may be required to continuously improve and stay focused on client satisfaction. If the control procedures are not established upfront, this could lead to failure of the initiative.

Inadequate pilot conditions

Some times a project team decides to launch a pilot before going full force on a process improvement initiative.

This is the case when the planned initiative is of a large scale and magnitude. Without having some solid results to back it up with, it is hard to deploy a new initiative across the board in a large organization. This is a common practice for initiatives that are planned for the entire organization. Executives wait to see how the pilot progresses and then they make decisions based on the pilot results.

What is not considered though is that a pilot could fail if the pilot conditions are not adequately thought through. In such cases, pilots should not be taken lightly. Pilot conditions should be close to reality and should be treated as a full fledged deployment. If the pilot for some reason is not conducted appropriately, this may lead the executive team to believe that whole initiative should be stopped. Since a lot hinges on how the pilot progresses, it is important to consider all aspects before launching the pilot.

Lack of resources

Last but not the least, lack of resources may also lead to failure of process improvement initiative specially if economic conditions are not suitable.

Even if everything was a go at the start of an initiative, the concept was well thought through and approved by the executive committee and execution was progressing well, sometimes, a project can get stalled due to resource issues.

In order to prevent this kind of failure, one should carefully consider budgeting and resources at the start of the initiative. Having said that, it is a changing world and sometimes, resource crunch happens and these conditions are outside of our control.

When this happens, one should halt the process and consider ways to continue what has already been done. In such a situation, one should make changes such that existing process can be maintained and controlled for maximum benefit.

This could be done by limiting the market segments or organizations in which the process is deployed. Instead of deploying to all organizations, perhaps the process can continue to exist in a limited set of organizations. If economy improves later, the process can be expanded to other organizations as well.

Process professionals should consider all of the above factors before launching their Process Improvement initiatives.

4 CHALLENGES FACED BY PROCESS PROFESSIONALS

In most scenarios, process professionals are introducing process in parts of the organization that are different from theirs. It is sort of a consultant, auditor or an outsider kind of role. There are therefore naturally many obstacles that process professionals need to overcome to succeed. The root cause of many obstacles come from these organizations that don't fully comprehend the importance or need for a process. They have little or no trust in a process consultant and his intention to help solve their business problems.

A common mistake amateur process professionals make is they underestimate these challenges.

In some cases, they ignore these challenges in the beginning without understanding the impact they can have on the success of the initiative.

Lets explore these obstacles in detail :

Not seeing the need for process

Most people in business world see process as unnecessary. They think they know what to do and how to do so why should a process be instituted? It could be true that perhaps 5% of people indeed don't need a process because they are highly organized individuals.

However most people are not that highly organized and would rather have a well defined process that is instituted for them to follow.

A process in a large organization is a necessity.

Whether it is a product organization or service organization, whether it is HR or Engineering or Sales, all business units need a certain well defined process to conduct business. Making target organizations see the need for process is a common obstacle that process professionals face.

Process is not seen in smaller start-up companies since there are fewer employees and they can run the business without establishing proper procedures. When these employees join larger companies, they find it hard to follow process or even see the need for process.

They see process as an overhead and unnecessary burden. The onus then lies on the process professionals to explain and make these people understand the need for process.

When a small startup grows exponentially, it becomes too large to be managed without a process. It is a challenge though for process professionals to make these people understand the need for process simply because they have always done things without a process.

Resistance to change

Resistance to change is the most difficult yet the most common obstacle that process professionals face.

If you are new to process field, you need to be aware of this obstacle before stepping into this field. If you can overcome this obstacle, success is yours!

But if this challenge is too big for you to handle, you should stay away and try some other profession. It is common human nature to oppose change specially if they don't completely understand why a change is required.

People are used to doing things in a certain way and they don't want to unlearn old ways and learn new ways. Other important point to understand is that any type of change takes time. A process professional needs a lot of patience to effect a change. Scale and magnitude of resistance can vary from situation to situation.

Sometimes it is a small resistance that can get resolved in a few days. Other times, the root of the resistance can be deep and it can even take months to overcome it. Overcoming resistance is the most crucial challenge for a process professional.

Small scale resistance can be seen when you are trying to implement a procedure within a small group of engineers. A bigger scale resistance can be seen when an organization wide initiative is being planned or rolled out.

Establishing credibility within target organizations

Since most process consultants are working to define a process in an organization different from theirs, they are considered an outsider to start with.

Lot of times they don't have their credentials established in the organization and that makes their job harder to do. Even if these process consultants are experts in their fields, they are still an outsider for the organization and therefore people have a bias against them. People take time to warm up to these consultants and don't accept their suggestions right away. It becomes extremely important for the consultants to establish good relationship with people within these organizations.

Process professionals may have developed credibility and are subject matter experts but building an emotional level credibility is the key to move forward. Without establishing these key relationship and credibility, success cannot be attained.

Art of influencing comes in handy to overcome this kind of resistance within a new organization.

If overcoming resistance to change is the most crucial challenge in the process, influencing is the most crucial skill needed by a process professional.

One should really master the art of influencing to succeed in this profession. It may appear as a difficult skill to acquire but it really is not. It takes some emotional intelligence and practice and knowledge of organizational politics to master this art. Another good way to learn this skill is to watch great influencers in action. You can always choose a mentor who is skilled at this art and learn by watching them.

Establishing right set of success measures

No process can be successful without accompanying success measures. Establishing the right set of measures for a process improvement initiative is one of the top challenges of process professionals. It is said "**what cannot be measured, can not be improved**". Therefore measurement is the key to improvement.

However the challenge lies in figuring out which measures are appropriate? Most of the time process professionals tend to establish complicated measures.

While complicated measures look good on paper and are perfect for publishing in journals, practicing professionals often find it hard to comprehend complicated measures. Even if they comprehend these metrics, they tend to forget them with time.

People find it hard to explain why these complicated measures trend in a certain direction. It is very important to establish metrics whose trends can be explained and utilized to make improvement. One should not establish measures just for the sake of it but also look at its practicality.

One should be able to explain why a measure goes up or down and what's behind the trend.

All of these factors make establishing optimal success measures a challenge for process professionals.

Finding change agents within the target organizations

Implementing an organization wide change takes a lot of effort and time. It is not a one person job to promote a change of this magnitude of scale.

Process professionals often recruit change agents within the target organizations to make the transformation scale. Finding the right change agents who have the passion to promote the change and believe in the purpose of the process improvement initiative is a tough task.

Establishing relationships in the target organizations and then locating enthusiastic change agents is one of the challenges. These change agents however make the task easier and are necessary for organization wide adoption of the initiative.

Art of influencing comes in handy to overcome this obstacle. Having the vision and passion for your initiative helps you recruit change agents and ignite passion in them as well. These change agents should be able to take the initiative and make it their own and run with it.

Securing sponsorship for the initiative

A process improvement initiative is not successful without management sponsorship.

Securing right executive sponsorship for the initiative is also one of the challenges faced by process professionals. This challenge is critical because you want a sponsor who believes in the initiative and is ready to champion it with you. This sponsor will obviously be in the organization in which you are tasked to implement the initiative.

If you are new to the organization, it can be a tough task to locate this sponsor. You will have to establish good relationships with people in the organization and understand the politics and the influencers in the organization. It is important to understand who is powerful and influential in the organization and who sees the need for the process improvement initiative. Although there are a number of good leaders in the organization, not all of them will be tuned to your initiative.

Not all of them may be facing the business problem that you are trying to solve. A good sponsor will see the business problem and having understood it, will stand behind the initiative to solve the business problem.

5 HOW TO OVERCOME RESISTANCE TO CHANGE?

Process Professionals more often than not come across resistance to Process Improvement and the change associated with it. This is the number one cause of concern for Process Professionals. Once you know how to master this challenge, you will have success all around. What you need to do is understand the psychology behind resistance in order to overcome it.

Why people resist Process?

Organizations and people within organizations are often seen resisting process. There is an inherent believe or bias against process – that it is not needed. It is also believed so because most often process professionals are part of a different organization or business unit and in some case consultants.

People often see process as adding extra overhead and don't see why they need to change the way they have been doing things. On a personal level, process may not be necessary to perform the job but at an organization level, it is very necessary for consistency and productivity.

I remember having this discussion with one of the software engineers when he was asked to follow a process to record the code review minutes during code review of his software program.

He kept on asking "what is the need to record the code review minutes?" He said "I always do code review and take care of comments but why do I need to record it?" That was his viewpoint and quite appropriate from his perspective, however what he didn't realize was that not everybody performed a code review and took care of code review comments.

How do you ensure that everybody does indeed review his code before it is checked in the source repository? Without introducing a process and guidelines that each and everyone ought to follow, it is difficult to expect a consistent outcome from all software engineers.

Without a proper code review and the action to address the issues found during code review, quality of software code suffered. This resulted in a number of software defects being found by the customers. This also led to poor perception of software quality resulting in poor business.

Poor quality perception forced the need for a process to be instituted to mandate code review. Code review of all the software programming code written in the organization is required to improve the quality of the software program. This new process ensured that not only code review was performed but also the issues uncovered during the review session were recorded. This process made it easier for managers and quality auditors to ensure that issues were addressed in time before customer release of the software.

In the above case, the resistance came from the engineer unaware of the big picture and simply focusing on his own practice. Many a times, the resistance is legitimate and has a reason for it.

As process professionals, we need to understand the root cause of the resistance and ensure that the right solution is applied to overcome resistance.

After the engineer was explained the big picture which was not obvious to him from his vantage point, he realized the need for the process and was happy to institute and follow it.

Resistance to process can come from different levels in the organization. Sometimes it comes from an engineer like the example mentioned above and sometimes it comes from a manager or higher executive level.

Given that both engineer and manager have their own set of objectives and goals that they are measured against, their motivations are also different. Sometimes a process might hinder with the goals of a manager at a personal level. He may realize that the process is indeed needed for the benefit of the organization. However, at a personal level, he may resist it because it will mean more work for him. It will mean he will need to get more organized and create more reports.

It might mean he will be held accountable for certain things that he was not being monitored for . This is not ideal behavior but is quite common in corporate world.

Similarly an executive has their own set of goals and agendas such as budgeting, staffing etc. If the proposed process hinders with any of these objectives or makes it harder to get staffing and budgeting, naturally they will resist the process.

Resistance from a managerial or higher executive level is harder to deal with as compared to say resistance from an engineer. In this case, help of senior management needs to be seeked.

Let us look at different reasons behind resistance to change.

Afraid of Risk

There is some amount of risk involved with any change and most people don't want to take risk. Risk is not an easy word or concept for most people to fathom with specially those that are conservative in nature.

However, what people don't realize is that there are risks and there are calculated risks.

Most successful leaders take calculated risks because they know without risk, there is no reward. Obviously one should not take unnecessary risks or risks for the sake of it. However calculated risks can bring about a lot of success.

Too comfortable to change

People are comfortable in a known environment. Although they know that existing process does not fulfill the requirements, they are comfortable following it and don't want to change the status quo. There is a certain comfort in doing things the way they have always been done and so why change?

If people don't trust the person initiating the change, this resistance becomes even more stronger. They start questioning the motive and integrity of the person initiating the change and come up with various reasons why no change is needed.

Emotional Resistance

Sometimes what you would see is emotional resistance. A process creator is emotionally tied to existing process that he created. In this case, he would not receive any ideas to change positively.

Process creator has put his heart and soul into the process. Emotionally speaking, he cannot let it change although on an intellectual level, he also realizes that a change is long due or will benefit the overall process. It is the pride or personal ego that comes in the way in this case.

Not Thinking outside the box

Some people simply fail to think outside the box. They follow what they have always followed and their thinking centers around it.

They don't get the big picture and they are looking at things from only their perspective.

When an process consultant comes in with a different perspective and perhaps a different set of experiences and offers a suggestion to change, they fail to see the consultant's perspective and therefore resist change.

Organizational Politics

Another reason of process resistance is politics of the organization. Understanding organizational politics is a tricky thing but very important specially for process consultants. Before embarking on a change initiative, it is important to understand who are the key players and what are their goals in the organization.

Is there any duplication in the organization? Are there any rivalries and one organization does things differently than other? Does one organization have a more powerful leader than the others? Which leader is more influential and why? Which leader will be more supportive of a change initiative and who will not be?

Techniques for overcoming resistance

Stating the obvious, when you come across resistance to Process Improvement, you need to think why and where the resistance is coming from? What is the source or root cause of the problem out of all the reasons stated above? This is the key to solving your problem. Often professionals struggle with this challenge since they are not looking at the root cause. They spend time in other solutions but don't quite address the root cause.

Once you have identified the root cause, you will need to address it. You will need to use tools to influence the people who are resisting the change.

For example, if the root cause of resistance is risk involved with the change - You will need to spell out all the risks and mitigation plan and pros and cons of taking the risk versus not taking it. When people see the pros and cons of taking the risk, they will immediately know whether taking the risk in that particular case is beneficial or not.

If the root cause of the problem is emotional resistance from the process creator, the worst thing one can do is not involve the process creator during the change discussion. This will increase their resistance. Instead what one should do is involve the process creator from the start. Take their input and talk through their journey of the process and humbly convince them that a change is indeed required. Remember intellectually they also know it is needed, the resistance is only emotional. What works best in this scenario is if you can make it sound like the change was indeed their suggestion. Have them feel the pride of suggesting a change much like they felt the pride of creating the process in the first place. If you follow this line of thinking, soon you will see they will be the biggest change agent for the initiative.

If the root cause of resistance is people not thinking outside the box, the solution has to be different. A lot of time will have to be spent on making them think like you, giving your perspective to them, making them see the world that is familiar to you but not to them. Pushing to make the change without having spent time to explain your perspective will hinder the process and will take longer to bring about the change.

When people are too comfortable in the current environment or way of doing things, the resistance challenge is a bit different. In this case, process consultant needs to do some research and show to these people various concrete examples of how a change brought positive results in other groups or organizations. Back your scenarios with concrete and indisputable data so the people resisting change can see for themselves and decide. Most process professionals often make the mistake of showing too much data or data that is massaged. Don't massage the data and show data that is indisputable. This will bring the best and quickest results to overcome resistance.

If organizational politics is the reason for resistance to your process improvement initiative, learn to play the game. In politics, the powerful always wins. In order to succeed as a process professional, you need to identify that powerful leader in the organization who is not only powerful but is also influential. You need to align with his goal and work in collaboration with him to further your initiative. If there is some politics involved, understand it first and then take advantage of your understanding. Lets take a look at an example of understanding and taking advantage of politics.

While I was leading a change initiative in an organization, I learnt that there were two rival business units who were always in healthy competition with each other. Each business unit wanted to show better results than the other.

One of these business units was in the process of adopting the change initiative I was leading but the other was showing resistance. Taking advantage of my knowledge of rivalry between these business units, I presented the case for the resisting organization to also adopt the change because the other organization was already adopting it. Realizing that they will be left behind, pretty soon they agreed to adopt the initiative.

Overcoming resistance to change is never easy. It takes time, persistence and perseverance but the reward is outstanding. In my career, I have found the most difficult task is changing people's mindset but it is also very rewarding.

What you need to know is the root cause of resistance, have the right tools to influence, get rid of the root cause and be persistent.

Know that you will succeed because you are addressing the root cause!

6 LEADING BY INFLUENCE

Leadership is a key success factor in any business or field. As a process professional, you are mostly working with other business units in the organization and working cross functionally, therefore leadership matters a lot. In order for a process professionals to drive process improvement initiatives, they need to be able to command respect and have integrity with their clients.

"Leadership both in terms of subject matter expertise and influencing skills are a key component of success."

When I took on the management role, I remember asking a senior leader, "What skills does one need to lead people they don't manage directly versus those they directly manage?" The answer I got was "You need the same skills". I was a bit disappointed by that answer, not because I knew I needed different skills but my gut said it has got to be a different.

I was thrown in a leadership role where I was leading people whom I didn't directly manage and so I learnt on the job. When you are managing direct reports, you have authority over them by default. You are responsible for writing their annual performance reviews. They are implicitly accountable to you.

However when you are leading people who are not your direct reports, they have no reason to listen to you or be led by you or be accountable to you. So the question is how do you lead them? It is difficult, in fact much more difficult than leading direct reports.

Here are a few things that help in leading without authority:

Subject Matter Expertise

You ought to be an expert in your subject and be respected for your craft. Process professionals are often consultants to the organization in which they are instituting the process. For them to be seen as leaders, they ought to be the expert in their field and have a certain amount of credibility.

Relationship Building

When you are new to the organization and want to be seen as a leader, you ought to put more emphasis in relationship building with your indirect team. You need to understand their needs and problems and be able to empathize with them. This might take some time and effort on your part but it is a very important piece of success of you as a leader and for the success of the initiative.

Having a purpose

In a direct team, there is a build-in glue between the leader and the direct reports. I have found that in an indirect team, what glues people together is the purpose of the task or project. If the process professional has a purpose which is solving the business problems and displays that through his actions, it brings credibility. It is not only important to have this purpose but also to be able to demonstrate through your actions.

Having a vision

Last but not the least, it is very important for a leader to have a vision and to be able to articulate it.

Visionaries make people follow them. It is almost like personal authority or personal power which draws people towards you. It is not only important to have this vision but also make people see the vision and be able to relate to it. This leadership quality is extremely important when you are creating something new. When something is a mere concept, you need to be able to make people see your vision and stand behind it.

Leading with Influence

As a process professional, the most important leadership skill is the ability to influence.

We have all either been influenced or have tried to influence others in the course of our work life. Who would you need to influence as a process professional? To start with, you would need to influence management and the leadership team about the initiative since they are going to be sponsoring it. You would certainly need to influence the people who would have to change and follow the process. You would also need to influence people to become change agents for you to deploy the process since you cannot do it alone.

Sometimes you intentionally influence others and at other times, you unknowingly influence them. The question is what is that key factor that helps you influence someone?

The number one quotient needed to influence successfully is your integrity and motive.

What is your motive behind influencing, no matter how much you hide or underplay it, sooner or later people will figure it out.

Does your motive to influence have a purpose? Is this purpose to improve the business process? Is this purpose related to solving the business problem or are you simply following a business process model and going through the motions?

A lot of times, your need for influencing can be different from a business purpose and it could be legitimate. However your influencing power in such scenarios is limited.

All successful process professionals with widespread influencing reach have one thing in common, the integrity and their motive tied to some business purpose.

Integrity wins you friends and keep them forever. Not only that, it helps build up your reputation in the business world. You would get more business from word of mouth and based on your integrity and drive to solve business problems.

7 HOLISTIC COMMUNICATION

Communication is important in any field or industry. As a process professional, communication is all the more important. When process professionals are introducing process initiatives in organizations, it is extremely important for them to keep all stakeholders informed about their plan, execution and results. Communication is important in every phase of the initiative right from the start till the end.

Communication milestones

First of all process professionals need to identify what are those important milestones, where communication is mandatory for the success of the initiative. Most of the time it depends on the initiative. In general, communication is very important in the beginning of the formulation of the initiative to get all stakeholders on the same page.

Once the initiative has kicked off, it is important to keep the stakeholders in the loop by providing periodic updates.

If you are running a pilot program, it is very important to communicate the progress and results of the pilot in order for stakeholders to decide if the initiative should move from pilot to full fledged program. Executive reports are generally done on a quarterly basis in the span of a fiscal year.

When establishing training for the process initiative, it is extremely important to get the word out and communicate to the right stakeholders and practitioners. Getting the training modules reviewed ahead of time for adequacy is important. Remember, the practice of the process will rely heavily on the training provided.

If there are any metrics that the practitioners need to be aware of, those should be documented well for them to understand their meaning and trends. When an initiative is about to be completed, it is again very important to communicate the expectation going forward and how the process team will control the process going forward.

Modes of Communication

Once you have identified the important milestones for communication, it is also necessary to figure out the modes of communication that you will be using. Different modes of communication work for different purpose and different audience.

For example, process training is best done in person. One should also record any live training so that it can be viewed by people who were unable to attend in person.

For milestones such as pilot completion, an email or newsletter is a good medium of communication. For executive communication, a well presented power-point is a good medium of communication. Process professionals should put some thought into the right medium for communication.

Target Audience

In addition to formulating the communication material and choosing the right communication medium, it is also important to identify the target audience for the communication. Some time target audience dictates the medium you are going to be using for communication as well as what material to use in communication. Target audience needs to be segmented based on their interests and ranks in the organization. For example communication about the process to engineers will be different from communication about the process to managers and executives.

Don't lump everything into one communication letter or presentation, it will confuse the audience and material will become irrelevant. It is very important to tailor the communication based on the target audience.

8 MEASUREMENT AND ITS IMPORTANCE

Anyone who has worked on a process will realize that measurement is a key component of any successful process. It is said that "**if we cannot measure it, we cannot improve it**". When we are implementing a process improvement initiative, we ought to think about success measures. We need to think about how to measure the progress of the process? Whether the process is indeed solving the business problems or not? A true success measure will indicate if the business problem is being solved or not.

Depending on the scale of the initiative, you could have short term and long term success measures. Sometimes the initiative is a long term initiative and end results will not show for couple of years. In this case, in order to measure short term success, one needs to have both short term and long term measures defined.

As far as measurement to support a process is concerned, it is not necessary to devise complicated metrics. In fact the simpler the measures, the easier it is for people to understand, follow and implement them. If the measure is not intuitive enough, it is not worth spending time on.

If the measure is hard to explain, it will be hard to understand and implement. It will be even harder to use for any kind of process progress. It is easier to implement simple metrics and trend them.

Metric trends tell a lot about the progress being made throughout the life-cycle of a process. Metric trends also give us feedback into the process. Trends tells us what is working and what is not. It is as simple as going to the gym everyday and measuring your weight before and after the month ends. Does your weight lower due to the exercise routine that you are following or does it remain the same? If the weight is lowering, you can continue with the same routine, otherwise you need to make adjustments.

Lets take a look at a process (Customer satisfaction survey) where simple metrics and trends can help. Most companies conduct customer satisfaction surveys to get feedback from the customers on their products or services.

As important it is to conduct the survey, equally important is to listen to their feedback and act on it. The whole idea of surveying customers is to make improvements and changes in the products or services based on customers' feedback. Assuming the company is making improvements and changes, how will they know it worked?

This is where simple metrics and trending will help. Let us say, a company asked a question in the survey and got a low rating for that question in one year. They then decide to make improvements. After making improvements, the rating for that question in the survey should improve. This is where customer satisfaction rating trends year to year will help the process. The trend will help identify the effect of changes made. An upward trend will imply lets continue with the improvement changes. A negative trend will have to be investigated in order to figure out what more needs to be done.

Measurement is the not the answer to each problem but it helps us identify the problem and point in a certain direction. Measurements should be used within the context. Often people make the mistake of quoting measurement without a context and when that happens, measurements can be mis-understood.

This is why sometimes people tend to dismiss statistics and measurements. It is very important when quoting a measurement that we also quote the context in which the measurement was taken. Such as if 90% of polled people said yes to a question, it is also important to state how many actual people were sampled. If the sample was large, it means a lot but if the sample was a small number, then it doesn't signify anything.

When decisions are being made, it is very important to understand the context of statistics and measurements and take business decisions accordingly. Many process professionals make the mistake of not stating the context of the metrics. They sometimes massage the data. This can lead to distrust amongst the audience. Such massaging of data should be avoided.

A common mistake audience does is when they look at measurements that are not favorable, they tend to dismiss them as mis-construed data or unbelievable data. They also at times, start distrusting the person presenting the data. One thing that always should be kept in mind is all data can be explained. It is the context that needs to be understood and the reason for why the data is unfavorable.

It is not important in business that we always have favorable data. What is important is to understand why the data is unfavorable and what can we learn from it?

If the data is unfavorable, isn't it an opportunity to identify problems and fix them? This is where process professionals when displaying data should always show the context on the same chart, slide or whatever shape or form the data is displayed in. They should also make the effort to explain the data and why it is trending in a certain direction.

Types of Metrics

When deploying an improvement initiative, there are two kinds of metrics one should look at implementing.

One set of metrics are **adoption metrics**. Adoption metrics indicate how much adoption have we had of the initiative across the board. This is the basic indication of how wide spread the initiative is. If there are any gaps in the adoption, this metric will help indicate it. These set of metrics are useful when the adoption is done on a large scale organization.

The second set of metrics are called **effectiveness metrics**. These metrics are very important to gauge how effective the process has been in achieving the business goal. Sometimes a process initiative may be basic hygiene and need not warrant an effectiveness metric if everyone agrees, it needs to be instituted. However, if we are not sure if the process is indeed the right solution, an effectiveness metric is very much needed and can be used as a tool to gauge progress or any changes required.

An example of adoption metric is keeping a count of how many groups in an organization have adopted TL9000 procedures.

TL9000 provides traceability between requirements and testing. Testing is supposed to get better with TL9000 procedures and is supposed to catch more defects. An effectiveness measure for TL9000 procedures would be check how many defects are escaping the testing phase. If TL9000 procedures are followed and requirements to testing traceability is done correctly, very few defects should escape. This is why this metric will tells us about the effectiveness of the process followed.

Metrics and Goals

A common strategy used by process professionals in an organization is to establish goals for metrics once a metric has been agreed upon for a process. Organizations tend to establish an improvement goal for the metrics such as 10% upward trend in a fiscal year. Goals are good to have, however there are some things we need to be careful about goaling. A metric should not be goaled unless and until it has been placed in practice for a long time and is well understood.

A common mistake people make is goal a metric too soon without understanding its trends. If a metric is given a goal to increase 10% in a year but it is not known which factors lead for metric to go up or down, it is a fruitless goal. It is simply a waste of people's time.

What happens if this is the scenario, people start gaming the system and come up with creative ways to raise the metric just to meet the goal. The whole purpose behind the process and metric is forgotten and all energy is spent on meeting the goal.

There can be nothing worse than simply trying to meet the goal without an understanding of how and why.

Dashboards

When we are talking about metrics, we have got to talk about dashboards. It is not enough to construct the success metrics for a process initiative, the metrics need to be visible in order for executives to take business decisions. The more visibility a metric has, the more power it has in influencing business decisions.

Dashboards are a very effective way to display metrics and their trends. In the process and business world, most decisions are taken based on facts and data, this is where a well constructed and consistent dashboard comes in handy.

There are various aspects of a dashboard that need to be kept in mind. First of all, it should be easy and intuitive to use. The metrics on the dashboard need to be clearly defined and how they are computed such that there is no confusion around the data. The data on these metrics should be refreshed periodically. The date when the data is refreshed should also be published on the dashboard. If the dashboard can also display pie or bar charts for easy usage, that helps in preparing presentation material for business meetings.

9 PROCESS CONTROL

Once a process is defined and adopted in an organization, it is very important to keep a tab at its progress and its practice. This aspect can make a difference in whether the process will have a long life or not. Lets look at ways to control a process effectively.

Establish effective training

Training should be established for new employees joining the organization so that they get familiarity with the process and know how to follow it. Adequate websites and resources should be available to those who are practicing the process and may have questions regarding what is appropriate and what is not. There may be mandatory aspects of the process and there may be some aspects that are optional.

All of this needs to be spelled out for the practicing teams. Appropriate training helps in controlling a process and getting the desired outcome.

Sharing best practices

Another thing a process team can do in order to control a process is share best practices around the process with new and upcoming teams. We often hear the phrase best practice in business process management.

"A best practice is an indication of higher maturity level of an organization."

Usually business processes are created by process professionals for business units to follow. Some business units will adhere to the minimal process and some business units improvise or add their own flavors to the guided process. It is often seen that these flavors turn into best practices. Other business units can then learn from these best practices and improve their products or services.

Lets look at an example of best practice for code review process of software programming module. Code review process calls out that each software programming code must be reviewed by peers, issues found in the software must be recorded and triaged and a time frame should be established for when these issues will be resolved. A mature organization will follow this process and might also add its own guidelines for designating an expert in each area of the software module. In addition this organization will mandate that when a software code is modified, that expert has to review the software code. So instead of having a generic peer review, the review is more tailored towards having experts in a particular area review the code. This flavor further enhances the quality of code review process.

Many organizations display best practices and these should be shared with peer organizations for broader impact.

If different teams are following the process and these teams don't communicate with each other that often, the responsibility of sharing best practices lies with the process team. Although guidelines for following the process are documented, there might be a team that has done an outstanding job in following the process and delivered excellent results. This team's best practices need to be shared with other teams so they can also learn and emulate.

Institute an effective audit program

There are internal as well as external quality audit programs put in place in organizations to monitor the progress and practice of a process. It is the responsibility of the process team to provide enough information to the practicing team so all the information is in their hands.

Things to look for in an audit from an auditor perspective are how well the process is being followed, general awareness of the process in the organizations and what are the trends of the metrics established and finally identify any gaps. All of these mechanisms provide feedback and are avenues for process professionals to control the process. This ensures the integrity of the process as well as takes care of any drawbacks that may appear after the fact.

Use feedback avenues to control the process

By keeping a good control on the process, if things change and a need for improvement arises, process professionals can act on it.

It is sort of a preventing measure just like there is prevention and there is cure for a medical problem. For the team that is being audited, if they understand and follow the process, there is no need to prepare for an audit, be yourself, be honest and show your work.

For example, while doing the audit of a software module, a gap was discovered in the software development process. The gap was identified in the phases of development cycle from the perspective of software developers. In this project, various software modules were being developed across different teams.

What was discovered during the audit was a gap in the phases defined for software development. The software developers were told to follow the sequence of Requirements phase, Design phase, Coding phase, Unit test phase and finally Code Review phase for the development of their software modules. Following these phases, they were told to hand over the software modules to the test teams for the start of the test cycle.

The test teams were supposed to test the software modules based on functional requirements. This sequence of phases worked perfectly fine for small scale software development, however it didn't scale well for large scale development programs specially those that spanned multiple teams.

The problem was that each team would do their own unit testing and then the software modules were handed over to test teams to test but no one took the responsibility of making sure that individual software modules worked well together.

Things were breaking apart as far as integration of software modules was concerned. The process team realizing this gap therefore introduced a new phase called "Unit Integration phase" in the sequence of software development cycle. This phase was introduced right after "Unit Test" phase so once an individual software module was unit tested, came the phase of "Unit Integration phase". Software modules were tested in unison by various development teams in the Unit Integration phase.

This made a huge difference in the quality of software and prevented a number of design issues to be uncovered by test teams. This also helped in cost reduction because when these issues were found by test teams it was more expensive to solve them versus finding them in Unit Integration phase and fixing them right away.

It is also the responsibility of the process team to communicate any gaps that they have identified in the practice of the process and what remedy they are suggesting to fill that gap.

This is similar to sharing best practices instead what the process team is sharing in this case is a gap and how to fill it. Communication is very important in the process field specially if a process is deployed organization wide. Communication can be in different forms, emails, newsletters, website, blogs, web conferencing or live presentation. This is another avenue to get feedback from practicing teams and understand their perspective as well.

10 WHEN ARE YOU DONE?

A common question all process professionals ask is "When do I know that I am done with a certain process?" Is there still a need for streamlining or not? Can I add more metrics to enhance this process? Can I add more training to help those that are practicing the process? Can I add more advanced procedures to this process? Can I add more process control?

The answer is pretty simple. **Stop when you think the business problem has been solved.**

There is no need to over engineer the process. Nobody wants more process then needed. What we need is enough process to solve the business problem and then it is time to move on to tackle a different business problem.

It is similar to the situation when you start watching a soap opera. You love the plot and you continue to watch it. Then there comes a point in the life of the soap opera when you think that the producers should call it quits. But they don't, producers instead add more twists and turns.

One should know when to gracefully exit. The same is true for process, we just need enough process to bring order amidst chaos, not more, not less.

I was myself involved in a situation where I had to make a difficult decision on whether I should further a widely deployed process. I had pressures from senior management and peers. They were all living with me the victory of deploying this process company wide. They were becoming greedy and wanted more success. I have to be honest, I went back and forth on this decision millions of times in my head on what should we do if anything about furthering this process.

We had introduced the process, we had piloted it, and then we got a company wide policy to deploy it throughout the organization. It was historical and we were all very proud of the accomplishment. We had developed metrics around the process and these metrics overtime were going to yield revealing and enlightening results. The results would help in making business decisions.

The issue at hand was should we goal these metrics or not? On one hand, goaling of metrics is a good idea as it gives the practicing people a bar to achieve. It keeps the momentum and builds visibility.

However, all metrics are not meant to be goaled. I thought hard and asked myself what is really the business purpose of this process and metrics? Would goaling the metrics help in furthering this business purpose or would it distract from it? The answer to that question was the key to my decision.

In this particular case, goaling would have distracted from the business purpose of the process and metrics. Had we goaled these metrics, people's focus would have diverted from following the process to simply meeting the goal.

They would have instead spent time in figuring out how to meet the goal and game the system instead of learning from the process. Hence, I decided not to goal the metrics. It took a lot of convincing that it was indeed the right decision specially for those who were looking to benefit from more visibility around the process. But it was time to move on to a different business problem.

Another key message in the above example is that one should always give priority to what is right for business and not personal or power goals. It may seem like a win in the short term if we achieve a personal goal, however it ruins our reputation as a process professional.

Credibility and reputation are very important for a process professional or consultant. This is because we are most of the times working for an organization different from ours. It is therefore very important for us to build that trust and credibility with our clients

and customers and always keep the business goals as
our first priority.

Appendix

This book is written from my practical experiences and is designed to help process professionals. This book will be very helpful to those that are new to this field or are contemplating taking process improvement as a profession. The scenarios faced by process professionals will be different in different fields, however the techniques taught in this book are transferable. These techniques are generic and can be applied to any field of business process.

A very easy to describe and simple methodology that I use to assess a new business situation is called L.E.S.S. (**Listen | Evaluate | Strategize | Systematically Execute**). You can follow L.E.S.S. methodology for Process Improvement in any business or field.

Four simple, yet very effective steps that provide maximum efficiency for any business process improvement.

Lets take a look at these steps.

Listen

The very first step is to carefully listen to your customers, stakeholders, clients and/or vendors. Use simple survey mechanism or informational interview process to understand how the exiting process works. By asking appropriate questions, you will gain a very good understanding of not only the existing process but most importantly the shortcomings in the process and opportunities for improvements. Throughout this phase, be open and listen carefully!

Evaluate

The next step in this methodology is to evaluate the inputs gathered in the Listen phase. Synthesize all the inputs from interview/survey mechanism and bucket them in few categories. Be mindful of where the information is coming from and who is providing the input. For e.g. a higher level executive has wider visibility and so perhaps his/her input might provide a wider perspective of the problem. Also look at any competitive data in the respective field to supplement the inputs gathered and draw your conclusions based on the information.

Strategize

Once you have evaluated the information & data and drawn your conclusions, you have most likely identified improvement opportunities and problems that need fixing. Plan and strategize on process improvement opportunities. A very important part of strategizing is to get appropriate management support for your strategy since no process improvement effort can be successful without adequate management support. Once you have gained sponsorship and management support, you can move on to the next phase.

Systematically   Execute

Finally, after identifying improvement opportunities and strategizing, you are ready to execute. However, it is very important to realize that execution needs to be done systematically! First of all, we need to set goals for the process improvement that we want to execute on. We need to then define appropriate phases of Process Improvement execution and start executing. While we are in execution phase, what is very important is to check against the defined goals and improvise if needed to keep track of the goals.

Use this methodology in your business field and have fun with it.

ABOUT THE AUTHOR

Aditi Chopra is a process consultant and an inspiring leader in the software industry. Aditi has led numerous process improvement and organizational change management initiatives in her career.